Superorganic Evolution

Superorganic Evolution
Nature and the Social Problem

An Essay by **Santiago Ramón y Cajal**
Winner of the Nobel Prize in Physiology or Medicine

Translated and edited by
Lazaros C. Triarhou, M.D., Ph.D.
Bodossakis Foundation Laureate in Neuroscience

CORPUS CALLOSUM
Thessalonica · Indianapolis

Superorganic evolution : nature and the social problem.

A Corpus Callosum book published by

Lazaros C. Triarhou, MD PhD
Professor of Neuroscience
University of Macedonia
Egnatia 156
Thessalonica 54636 (Greece)

Ramón y Cajal, Santiago, 1852-1934, author.
    [Evolución super-orgánica : la naturaleza y el problema social. English]
    Superorganic evolution : nature and the social problem : an essay by Santiago Ramón y Cajal, winner of the Nobel Prize in Physiology or Medicine / translated and edited by Lazaros C. Triarhou, M.D., Ph.D.
    p. cm.
    Includes bibliographical references.
    ISBN-13: 978-1-5309-9962-0
    ISBN-10: 1-5309-9962-6

    1. Psychology–Philosophy. 2. Art–Philosophy. 3. Education–Philosophy.   I. Triarhou, Lazaros Constantinos, editor, translator.   II. Title.

Printed in the United States of America

Cover photos.—Sunset over the Pacific Ocean (obverse) shot from the International Space Station (credit: NASA). One of the last portraits of Cajal at 80 years (reverse) made by "Campúa" or José Luis Demaría López (1870–1936).

The Corpus Callosum logo is based on a lithograph by the neurobiologist Christofredo Jakob from his *Icones Neurologicae* (1897).

CONTENTS

FOREWORD

The Spanish histologist Santiago Ramón y Cajal (1852–1934) shared the 1906 Nobel Prize in Physiology or Medicine with the Italian pathologist Camillo Golgi (1843–1926) in recognition of their work on the structure of the nervous system. By putting a silver nitrate staining method, which had been developed by Golgi in 1873, to excellent use, Cajal elucidated the minute structure of the nervous tissue [DeFelipe, 1999; DeFelipe and Jones, 1992].

The driving force of laboratory experimentation was only one of Cajal's talents [Romero, 1984]. What makes him a genius is the tenacity of many of his scientific and social ideas, which were ahead of his time [Ramón y Cajal Junquera, 2002].

An acute observer of the literary and scientific scene as well as of the roiling social life of the avant-garde, Cajal emerged in 20th century Spain in a further role as philosopher and educator. He even came to stand in some sort as a symbol of national cultural rebirth [Sherrington, 1935]. One notes with delight the consistent references by Cajal not only to the findings of research scientists, but also, to the contributions of great philosophers and thinkers [Pasqualini, 1999], proving his erudition, and rightfully earning the titles of "Cervantes of Science" [Cannon, 1949; 1951] and "Don Quixote of the Microscope" [Williams, 1954; 1955].

Among several key 19th century neuroscientific concepts, it is the broader context of *Naturphilosophie* [Clarke and Jacyna, 1987] that becomes evident in Cajal's essays. One detects traits of the Romantic biology and medicine that was current among natural scientists during Cajal's formative years, in interplay with the convictions of physical-chemical reductionism. Or, in the words of Sir Charles Sherrington, "the simplicity of Cajal's ways and ideas illustrates how singular a mixture he presented of old-time ways and ultra-modern science" [Cannon, 1949].

Cajal's aura exceeded the boundaries of the Hispanic world. The British pathologist William H. McMenemey [1953] remarked: "The scientist will be remembered not only as the greatest of neuro-anatomists, but also as a writer of distinction, a pathologist, a philosopher and a patriot. He secured for Spain a place of honor in a rapidly expanding world of medicine, establishing a school of histology which revolutionized the methods of approach to the problems of neuroanatomy, and influenced the contemporary worlds of physiology and neuropathology. For him the horizon of histology was limitless and all might share in the work and the satisfaction it gives."

The material of the present booklet is a revised extract from my larger volume of Cajal's essays on psychology, art and education [Triarhou, 2015]. Those critical essays, time-transcending evergreens, portray "the catholicity of his interests" [Courville, 1953]. They shed new light on the breadth of Cajal's mind and bespeak the varieties of the empirical endeavors of the human brain in behaving and knowing. Bringing together imagination, passion and ideas, they help us to understand or rethink what we thought we already understood about Don Santiago's own brain.

The aim of the book remains similar to that exemplified years ago by Craigie and Gibson [1968]: "To bring together the literary rather than the scientific works of this great man. They represent a commentary on life as lived anywhere at anytime."

Lazaros C. Triarhou
April 2016

ENRIQUE LLURIA

EVOLUCIÓN
SUPER - ORGÁNICA

(La Naturaleza y el Problema Social)

CON PRÓLOGO
del
Dr. D. Santiago Ramón y Cajal

Enrique Lluria

The Cuban urologist Enrique Lluria (1863–1925) and his book
Superorganic Evolution, prefaced by Ramón y Cajal [1905].

EDITORIAL ANNOTATION

Prólogo.—Evolución super-orgánica: La naturaleza y el problema social [1905]

Enrique Florencio Lluria Despau (1863–1925) was a Cuban physician, turned sociologist. One of the first Cuban materialists, he was born in Matanzas, and enrolled to study medicine at the University of La Habana in 1882. He moved to Barcelona, where he completed his medical studies in 1889. Lluria next trained in urology (1889–1893) at Hôpital Necker in Paris under Joaquín Albarrán (1860–1912) and Félix Guyon (1831–1920), and made notable contributions to that field [Sagué Larrea, 2012].

In 1893 Lluria moved to Madrid, where he opened a clinic of urology and met Cajal. The two men became friends and collaborated in neurohistology. Eventually, Lluria surrendered to sociology and authored several monographs. Lluria frequently gave lectures at the *Centro de Sociedades Obreras* (Center of Workers Societies), the antecedent to the *Casa del Pueblo* (People's House) in

Madrid, and was affiliated with the periodicals *Vida Nueva* (New Life) and *La Revista Socialista* (Socialist Review).

In 1905 Lluria published his first sociological book, called "Superorganic Evolution," where he unraveled his views on the relation of sociology to the natural sciences; Cajal contributed the preface [Ramón y Cajal, 1905]. A sequel followed, called "The Humanity of the Future" [Lluria, 1906], as a natural complement of the initial work; in that sequel, the epilogue was written by Charles Malato (1857–1938), one of the best known pre-war publicists of anarchism in France. Both works have been widely reprinted.

Lluria lost his first wife, Clara Iruretagoyena, six years into their marriage, after having had three children. He was remarried to María Vinyals, a niece of the Marquis de Mos and heir of the Sotomayor Castle in Pontevedra. In 1909–1910 Lluria built a sanatorium by the castle, which was soon boycotted by the news that it had become a gathering site for socialists. Although Lluria considered himself a socialist, he did not formally join the Spanish Socialist Labor Party *(Partido Socialista Obrero Español)* until January 1915, prompted by the horrors of World War I. In July 1918 he left the Socialist Party on grounds of disagreement over class struggle and strike abuse, which he thought should be substituted by educating the workers. In 1919 the couple returned to La Habana. Lluria died in October 1925, at 62 years of age. Earlier in that year, he had moved to Cienfuegos and had opened an office at Calle San Carlos 161 [EcuRed, 2012].

Cajal's preface to *Evolución Super-orgánica* became popular with the anarchist and the socialist press. It was reprinted in the weekly magazines *Tierra y Libertad* (Barcelona, no. 55, 22 March 1911) [Zambrana, 2009], *Vida Socialista* (Madrid, no. 127, 14 July 1912) [Ramón y Cajal, 1912b; 1996], and *El Porvenir del Obrero* (Mahón, no. 340, 30 January 1913) [Núñez, 2005]. The text attests Cajal as an ardent supporter of social or "superorganic" evolutionism.

An English translation of Cajal's preface was rendered by Rachel Challice and Daniel H. Lambert for the British edition of Lluria's book [Ramón y Cajal, 1910]. The present translation was reworked to match more closely the original Spanish text [Ramón y Cajal, 1905].

Preface to Enrique Lluria's "Superorganic Evolution: Nature and the Social Problem"

by *Santiago Ramón y Cajal*

DR. LLURIA, having granted me the courtesy of considering me of competence in sociological matters, amiably invited me to express my opinion on his present book, dedicated to the study of the anthropological causes of the so-called social question.

Such a request puts me in great difficulty, as I am not conversant with the science created by Auguste Comte and developed by Herbert Spencer; I have occupied myself very little, or, better said, I have not had time to occupy myself with the moral and intellectual evolution of humans, considered in its relations with Society and the State. A worker bee of the great human hive, I have mostly confined myself to gathering honey in the garden of Nature, in order to build my small and individual cell, leaving others, with the eagle's vision and synthetic genius, to trace the perspective and find the philosophy of the common work, marking the future routes of the human swarm.

However, since in this case my silence would constitute an unmerited gracelessness, I am going to respond to the honorable invitation, explaining, with neither dogmatic airs nor *arrières pensées*, my intimate impressions on the doctrine developed by Dr. Lluria, and the solution, still quite remote, of the terrifying social problem.

I am entirely in accord with the critical part of the present book. Its author is right in declaring that contemporary Humanity, the *superhuman organism*, as Dr. Lluria calls it, has contemptuously severed herself from Nature, occasioning this systematic and perpetual violation of evolutionary laws with irritating inequalities and torturing pains and miseries.

The social human of today, adulterated by the morbid adjustment to the capital, is becoming an estranged mixture of civilization and barbarism. He thinks and feels, it seems, as a Christian, but he acts in the manner of a citizen of the ancient aristocratic and inhumane Republics. The sphere of his intelligence has increased and the sphere of his will has decreased to the same degree.

Increasingly more refractory by the day to the sense of justice, the current society offers us the sad and paradoxical spectacle of a world upside-down; at the top, vice and laziness sit enthroned and venerated; at the bottom, struggling with hunger and pain, the toilers and the useful, i.e. the brains that, as Spencer would say, have better adapted, sharpened by hard necessity, sovereign sculptor of the nervous clay, the internal to external dynamic relations. Hence the inevitable decadence and weariness of the human race; whereby the superiorly adapted organisms, consumed by overwork and misery, fall into sterility or leave behind a ruined posterity, decimated by infections; whilst, on the contrary, the drones, the inadaptable, the poor in spirit, surfeited with pleasures, incubate a robust offspring, perpetuating thus the dead weight of the social machine.

Therefore, for civilized man, the principles of

the selection of the fittest do not promote the cast of the best in life's struggle; instead, as Dr. Lluria shrewdly notes, adaptation is adjusted to an artificial extra-organic condition, quite unknown in the rest of the animal world, and an inexhaustible cause of aberrant weariness, retrogression and organization; for example, the acquisition and enjoyment of capital with the sole end of guaranteeing the perennial leisure of the few and the incessant increase of the parasites of labor. So that the human type, perpetually oscillating between misery and abundance and between anemia and plethora comes to being something extraneous and incomprehensible; a species of the insane, afflicted by the rare mania of imposing hunger upon others in order to procure for himself the extreme voluptuousness of committing suicide out of satiation.

Agreeing with the author, I reckon that the only capital that is anthropologically legitimate is human organization and the forces of Nature, factors of production which cannot march in consonance with justice and the law of evolution, unless collectively sustained and administered. The earth for all, natural energy for all, talent for all: this is the desirable division of future society. It then becomes urgent, as Dr. Lluria declares, to reintegrate man into the laws of evolution, by giving back the capital, which has been sequestered to the enjoyment by a few, to the common cultural heritage, continuing, in the end, as Cánovas del Castillo would say, the biological history of the human race, wearied of the selfishness and injustice of three thousand years of civilization.

But is this possible? In the event that this does not represent a beautiful and flattering dream, how will it be realized? Will the mighty human, piously expropriated for the common good, resign to mediocrity? Will not perhaps atavisms of the dethroned autocrat and the secular instinct of the enslaved ant be pulled away from his heart, his wrathful hands armed? And if we have to suppress by force these perilous nostalgias of untamed leisure, will not the future society be threatened by new wars of classes, with the consequential overwhelming expenditure of soldiers and cannons, and the irremediable overwork of the best? And even in the seductive hypothesis whereby peace is re-established and the world is transformed into a vast workshop, where moderation and love preside, how can one avoid the sexual instinct, acting without brake or prevision, from flooding into life millions of hungry mouths, an overwhelming charge upon society and a constant danger to collective peace? And if eventually the thesis of Malthus proves true! What will our future statesmen do with the excess of population, when, with America and Africa crammed with European emigrants, are left with no virgin soils to plough and mines to exploit?

And turning the attention to the march of civilization itself, will not the *aurea mediocritas* to which socialism aspires, not enervate the faculties of the spirit, undermining the energy for scientific inquiry? Will not collective capital be timid, and lack the dash, on romantic and savior occasions, of the individual capital? Will glory, the passion of philosophical and scientific genius, prosper in the gray and subdued environs of the collective comfort? With justice banished, will not perhaps the best spring of

Humanity's mental evolution have ceased to function? Pain is a great moulder of the will, and a promoter of heroism! When misery and disgrace are reduced to a tolerable minimum, will not the sublime abnegation of heroes and the potent genius of redeemers decrease in equal proportion?

To all these torturing doubts and questions Dr. Lluria answers with a doctrine highly sympathetic and inspiring.

This is how I interpret it.

Current production, the work of a hungry and uneducated minority, is deficient with relation to the necessities of the race. Divorced from the natural laws, our brain evolves with devalued and sparse fruits. And as an inevitable consequence of the nutritional poverty and the rigors of overwork of the majority, moral and physical pain is produced, in tandem with physiological misery, degeneration of the species, and, in the moral sphere, hatred of the classes and a discontent for life.

But such a deplorable state of affairs cannot be eternal. A time will come when science will illumine the mind and elevate the heart. And then, when, having banished the fetish cult of the capital, humans will become incorporated into the laws of evolution; when, having scrutinized and exploited natural forces, the Cosmos works for us, setting in motion an endless number of machines and manufacturing articles at ridiculous prices; when, having discovered the secret of chemical syntheses, the engineer of the future elaborates—without competing with the earth—the seed, the gluten, the albumin, the sugar and the fat, utilizing in effect the living force of the solar radiation or some form of natural energy; when leisure well earned permits the universalization of science and art, and everybody can taste the ineffable harmonies and beauties that palpitate in Nature's floor; when, finally, redeemed by solidarity and love, we all feel within us waves of one and the same vital current, sister-cells of the same body…, what meaning will attach to the words "rich and poor," "master and slave," "happy and unfortunate"? What will it then matter that love multiplies the species with no end, or that a blazing sky and a barren earth do not haggle over their gifts? Full of energy and alertness, to react against all kinds of cosmic accidents, the human brain, exalted by its fidel adaptation to the mechanism of the world, generously offers us new and salvaging inventions. Ours too will be the Treasure of the inextinguishable solar bonfire, which science, emancipated perhaps from our ancient and fatigued *nutrient*, the earth, will be able to turn and adapt to the production of glittering fruit and golden harvest. Who fears the exhaustion of solar force, the movement of the winds and the seas, the falls of the cordilleras, the sovereign power of thought?

Proud and encouraging ideal, that may perchance one day be converted into a vivid and palpitating reality!

Let us believe that its advent is at hand; because in this low world only what is actively believed and expected becomes realizable...

Setting aside the doctrine and the luminous horizons that its author reveals to us upon evoking, with a prophetic vision, the future society, disdainful of the individual capital and addicted to the cult of Nature, there are in this book many suggestive ideas and concepts which, even separated from the fundamental thesis, have proper value and brilliance, like jewels set in an artistic crown.

One of them is the assimilation of life to a rhythm, to a system of waves, comparable in principle to the palpitations of ether or to the most complete order of relations marked in the tables of Mendeleev and Sir William Crookes.

At first sight, the idea appears obscure and even difficult to conceive; but meditating upon it, one discovers luminous facets and most interesting points of view.

In brief, life represents a complex system of forces, of vibrations in ascending progression. Like an orchestra successively reinforced, the organization begins with the monorhythmic note of the infusoria and ends with the grandiose symphony of the mammal, in which millions of cellular voices collaborate. And when the volume of the organic orchestra reaches its climax, once again rises the chanting *refrain* of the germ, i.e., the simple cadenzas of the ovum, starting from which the melody develops with a *crescendo*, ever more complicated, until arriving again at the fullness of the musical modulations and motifs of the adult organization.

In no organic apparatus do we find this rhythmic character in a greater relief than in the cerebral instrument. Our spirit is nourished by waves gathered from all parts of the Cosmos, and its principal mission consists in classifying, combining and reflecting them, with reference to their origin. Perception, idea, spoken word, even muscle contraction, what are they, in the ultimate analysis, but palpitations of heat, of light, of chemical energy, of electricity, et cetera, transformed, refined and converted into other palpitations, more sophisticated and spiritual? Like a lens of singular virtue and power, our nervous system gathers all the murmurs and shivers of the world, in order to condense them, at times into the splendid spotlight of an idea, and at times into the flame of will and passion.

If considering the animal series as a chromatic scale—like a symphony performed by the natural forces, which, after forming the brain, play upon their nerve fibers like the wind on the harp—is an interesting conception, no less so is the attempt to explain the inheritance of acquired qualities by the trophic influence of the nervous system.

Certainly, the attempt is somewhat premature. We lack anatomical and physiological data to prove how an organ perfected by adaptation to the conditions of its environment can influence the brain to such an extent that the latter, in turn, modifies the germinal cells. Nor is there a sparsity of scholars, like Weissmann, who roundly deny the transmissibility of acquired characteristics, imputing all to the hazards of variation and natural selection. But, in the end, if the arduous problem has not at the moment met with a total solution, it is something to know that our ideas and feelings influence, by means of the great sympathetic system, upon the nutrition of the glands and the molecular architecture of the germinal cells. In every way, the proposition is tempting, and if science ultimately confirms this principle (the action of the nervous system upon the molecular arrangement of the nucleus and protoplasm), the neural theory of the inheritance of acquired qualities will replace the arbitrary hypotheses of Darwin, Haeckel, de Vries, and others, on such a most interesting problem.

Full of enthusiasm is also the hymn that Dr. Lluria chants to the indefinite perfectibility of the brain, this eternally young viscera that we all possess. A slave, firstly, of the cosmic forces that sculpted, with dolorous minuteness, the Daedalus of its associative pathways, the human brain is destined to become one day the tyrant of that same natural energy to which it owes its emergence. It is certain that the senses, narrow windows of the soul, have broken the continuity of the scale of ethereal vibrations, obliging us to select only the most useful for the rise and prosperity of the species; but equally so, by a wise compensation, our cerebral cortex, exquisitely plastic and creative, has succeeded in filling with ideas and inventions the gaps in the diminishing sensory register. What are the instruments of science, the microscope and the telescope, the galvanometer and the photographic camera, the screen of the fluoroscope and the resources of analytical chemistry, but complementary retinas and Corti apparatus, perceiving from distance, by virtue of which the human ingenuity, by emending Nature, enters into a possession of all palpitations of the cosmic energy?

And I conclude; because it is not my intent to sully with importune commentaries the diverse and attractive themes that, with an abundance of erudition and sound criticism, Dr. Lluria unfolds in his lovely work. It is written, I might add, clearly, brightly, suggestively, and with a valence of thought and a serenity of judgment which many gleaming writers in philosophy and sociology would yearn for.

S. Ramón Cajal

1905

BIBLIOGRAPHY

Cannon, D.F. (1949) *Explorer of the Human Brain: The Life of Santiago Ramón y Cajal (1852–1934)*. Henry Schuman, New York.

Cannon, D.F. (1951) *Vida de Santiago Ramón y Cajal: Explorador del cerebro humano* (traducción de A. Folch y Pi). Biografías Gandesa – Exportadora de Publicaciones Mexicanas, México.

Clarke, E., Jacyna, L.S. (1987) *Nineteenth-Century Origins of Neuroscientific Concepts*. University of California Press, Berkeley.

Courville, C.B. (1953) Santiago Ramón y Cajal (1852–1934). In: Haymaker, W., Baer, K.A. (eds.) *The Founders of Neurology*. Charles C. Thomas Publisher, Springfield, IL, p. 74–7.

Craigie, E.H., Gibson, W.C. (1968) *The World of Ramón y Cajal, with Selections from His Nonscientific Writings*. Charles C. Thomas Publisher, Springfield, IL.

DeFelipe, J. (1999) Cajal, Santiago Ramón y. In: Wilson, R.A., Keil, F.C. (eds.) *The MIT Encyclopedia of the Cognitive Sciences*. MIT Press, Cambridge, MA, p. 98–9.

DeFelipe, J., Jones, E.G. (1992) Santiago Ramón y Cajal and methods in neurohistology. *Trends in Neurosciences* 15: 237–46.

EcuRed (2012) Enrique Lluria Despau. *Enciclopedia Cubana*. Accessed from www.ecured.cu on 3 November 2012.

Lluria, E. (1906) *Humanidad del porvenir*. Publicaciones de la Escuela Nueva – La Neotipía, Barcelona.

McMenemey, W.H. (1953) Santiago Ramón y Cajal, 1852–1934. *Proceedings of the Royal Society of Medicine* 46: 173–7.

Núñez, D. (2005) Anotaciones a un artículo inédito de Ramón y Cajal. *Revista de Hispanismo Filosófico 10:* 111–6.

Pasqualini, C.D. (1999) Cien años después en investigación científica. *Medicina (Buenos Aires) 59:* 798–800.

Ramón y Cajal, S. (1905) Evolución super-orgánica; Prólogo. In: Lluria, E.: *Evolución super-orgánica (La naturaleza y el problema social).* Librería Española/El Siglo Nuevo, Barcelona – Librería de Fernando Fé, Madrid, p. v–xii.

Ramón y Cajal, S. (1910) Preface. In: Lluria, E.: *Super-Organic Evolution: Nature and the Social Problem* (translated by R. Challice and D.H. Lambert). Williams & Norgate, London, p. vii–xviii.

Ramón y Cajal, S. (1912) La sociedad del porvenir. *Vida Socialista 5:* 3.

Ramón y Cajal, S. (1996) La sociedad del porvenir. In: Moral Sandoval, E., Martín Nájera, A. (eds.) *Vida Socialista*, tomo 5 (Edición facsimilar de la revista publicada en los años 1910–1914, no. 103–129). Mainer Til Editores, Barcelona.

Ramón y Cajal Junquera, M.Á. (2002) Cajal en Barcelona: Santiago Ramón y Cajal y la hipnosis como anestesia. *Revista Española de Patología 35:* 413–4.

Romero, A. (1984) *Fotografía Aragonesa/1: Ramón y Cajal.* Diputación Provincial de Zaragoza, Zaragoza, p. 82–5.

Sagué Larrea, J. (2012) Enrique Lluria Despau, notable urólogo de origen cubano que ejerció en España. *Revista Cubana de Urología (La Habana) 1:* 19–22. Accessed from www.revurologia.sld.cu on 3 November 2012.

Sherrington, C.S. (1935) Santiago Ramón y Cajal 1852–1934. *Obituary Notices of Fellows of the Royal Society 1:* 424–41.

Triarhou, L.C. (2015) *Cajal Beyond the Brain: Don Santiago Contemplates the Mind and Its Education.* Corpus Callosum, Indianapolis – Thessalonica.

Williams, H. (1954) *Don Quixote of the Microscope: An Interpretation of the Spanish Savant Santiago Ramón y Cajal (1852–1934)*. Jonathan Cape, London.

Williams, H. (1955) *Don Quijote del microscopio: Una interpretación del sabio español Santiago Ramón y Cajal (1852–1934)* (traducción de P. Abelló). Taurus, Madrid.

Zambrana, J. (2009) *El anarquismo organizado en los orígenes de la CNT (Tierra y Libertad 1910–1919)*. Centre de Documentació Antiautoritari i Llibertari, Badalona – Buenos Aires, p. 56, 627. Accessed from www.cedall.org on 3 November 2012.